Designer Bookbinders in North America

June 2000 – October 2001

Almost thirty years ago, Designer Bookbinders held its most ambitious exhibition. It toured North America and, on its return to Britain in 1972, it was shown at the Victoria and Albert Museum in London. Quite by chance, I saw the exhibition on the very last day. Having served an apprenticeship as a trade bookbinder, I found these designed bindings a revelation, and it proved to be the day fate altered the course of my life. It feels extraordinary to me that, as we embark on this second major tour of North America, I am now president of Designer Bookbinders. I could certainly not have imagined this all those years ago.

Many whose work inspired me in that exhibition are now colleagues and old friends. They are still active and their current work is, I believe, even more ambitious and exciting. In contrast is the work of our new Licentiates, some of whom, as I write, joined us just a few weeks ago. Combined with the generations between, the spread of work is, as it has ever been, very wide-ranging. For me this has always been the strength of our society: the diversity of approach to the medium, the only demand being that the quality and integrity of this work is of the highest standard.

David Sellars

president of Designer Bookbinders

A new art of making books

We are asking the question: 'Is there anything specific or characteristic about British Bookbinding[1] today?'

With so much ease of communication and an international interchange of ideas via the Internet, together with all the variety of techniques and styles visible on the international book arts scene, this is a very involved question to answer. The collection of books shown in the accompanying exhibition and in this catalogue could give more answers than words can indicate, but a great deal is lost to the spectator just because he *is* only a spectator. That the books are confined to 'glass' cases (by the necessity for security) means that the hidden aspects of the work are lost to the viewer. These hidden aspects include not only the physical structures of the books themselves but also the inner workings of a mind in response to what authors are saying within the texts, or the response to the book as an object made from a variety of materials.

We hope in this introduction to point to some of the missing elements and to specifically British approaches which are included within the body of this exhibition.

A survey of how some British bookbinders approach the modern medium can be synthesised into what aspects of the book enthral them, and what it is about making books that gives them satisfaction and drives their endeavours. These could be legion, but there are some pointers which come over almost as a consensus, not all openly expressed but none-the-less apparent by matching words to deeds, after discounting the lip-service. There are four aspects to making an expressive book work; all of which are demonstrated in the exhibits:

1. The *craft* and techniques of making; the physical skill.
2. The organisation of materials, structures and visual/tactile elements; the *designing* aspect.
3. The *content* of the image/design; the artistic ingredient.
4. The *level of the content*, combining all three basic factors in a unified whole, but adding special significance, that is, 'meaning' beyond the sum of the parts. (To be all embracing this must be influenced by the *level* of the content of the chosen text.)

Each area (taken as a separate speciality) can have its own 'meaning' for a maker. Most makers are lit up by one or more of these factors. It is self-fulfilling for them and satisfies their needs as makers. We will examine these factors below.

With the inevitable exceptions to rules, there is a British 'attitude'. Individuality of expression is one of its distinguishing marks, but there is an interplay of factors influencing styles and techniques; ideas are always being bounced off each maker on to others. British makers have advanced the search for new and exciting book structures and new techniques, stressing the importance of sound materials and book mechanisms, probably more than in any other country[2]. British bookbinders have no *atelier* system and therefore (with occasional exceptions) undertake all the aspects of book making as individuals. This encourages the exploration of new ideas, and is having a positive effect on the work in other countries, notably through publications like *The New Bookbinder* journal.

Britain is generally noted as a rather literary nation, represented by some of the world's greatest authors and poets. Most of those in Designer Bookbinders use these authors and poets as a basis for their inspiration, although responses are highly subjective and it is not easy to 'tell a book by its cover'. Sometimes the design contradicts the feelings generated by a text. If the title is used on the exterior maybe the design will 'illustrate' that literally rather than the 'plot' of the text. Most of us avow that we are relating visual appearance to the contents of the text and this actually means matching up the feelings generated by both.

These two aspects of a book (outside and inside) are often very hard to co-ordinate when one can only look through a glass case. Should one be able to tell where the text is heading by looking at the outside? Is the imagery or design helping or extending the text in a meaningful way? The exhibition-goer cannot look at the illustrations inside the book, unless the book is displayed open with a sample page visible. Each book is static like a butterfly pinned out in a lepidopterist's tray. Unless the text has been read there will be few clues as to what the book maker is driving at with the design. So does this mean that it does not matter what one uses to decorate the outsides of a book? Not to the casual observer certainly. The book will be looked at simply as a more, or less, interesting visual, perhaps three-dimensional, artefact, or a spectator will be so drawn to the demonstrable skill of the finish that he will not notice much else.

So how does this question of anonymity of the exterior design affect the attitude of the creative maker him- or her-self when he has admitted that it is his aim to have a definite relationship with what the author is saying? Surely this is important as *the* challenge of this medium: that it is a literary and visual three-dimensional art? Many of those questioned say that this *is* the challenge and that it is some part of, or even the total apparatus of, the book which stimulates their creative endeavours. Why cannot there be an inherent unification of all the aspects? Why leave some out? In the new art of making books this is an important question. Is what the author is driving at coming across in the way the exterior is treated, or have the so-called limitations imposed on another's creativity by a 'subject' defeated the maker? The verbal 'explanation' of a work lies within the text. Text and external treatment should be sufficient to 'read' the whole work. The work must stand on its own, but cannot be 'read' entirely by looking at the outside.

There are some makers here who actually indicate by their work that they are not too interested in relating to specific texts, but they may be interested in structural mechanisms and in book action, or simply in creating harmonies with the materials which are used to construct the book as an aesthetic object, or they are interested in creating a 'slick' finish. Is it necessary then that a text is involved?

The use of a ready-made text is not the only option in the making of art objects from books. Some of our most creative makers work on the contents of the text block themselves (or alter it) and this enables a more holistic and stylistically related work, with more control over materials and procedures. The artist's book is now on the agenda of the 'fine' bookbinder[3]. Those who feel no particular desire to create texts, visual or verbal, select those texts which, had they the skills of the great writer, they would have liked to have written themselves. The maker has a particular view of "life, the universe and everything". So why not select texts which seem to be in accord with, or moving in the direction of, that world view? The selected text can be used as a starting point in extending the *'book-art-maker's'* dialogue with it. Visual artists have their own conceptual viewpoint and try to express this in various media. Why should the *'book-art-maker'* be any different in his or her approach?

With craft objectives in mind, a group of works by an individual will appear 'much of a muchness', and the bindings' visual designs could be interchanged between texts without loss of visual and tactile appeal. There are other objectives which British book makers pursue, often related to the artistic objectives: investigations concerning the way books open, the fallibility of traditional hinging methods, the necessity or otherwise of sewn endbands, squares, back-linings, tubular hollow backs, sewing structures which will not damage the print carrier, the manipulation of leather in new ways. These are of course investigated by other cultures but British bookbinders and conservators have pioneered many of them and news of the innovations spreads world-wide.

To relate to the spirit of a text, an image need not be illustrative of either the title or the text, but questions of harmony and perhaps practicality must be considered. A book is not a painting on a single surface, it is a mobile articulating solid containing the work of another artist and so should be able to say more by its changing and alternative surfaces than a static wall-piece. Today British bookbinders feel that decoration for its own sake does not sit comfortably on certain works of literary merit. Literature is after all an art in its own right and to celebrate the individual expression of a respected author with unrelated decoration seems to some to be a kind of insult, even sacrilege, tantamount to ignoring the *raison d'être* of that medium. Why use a book as a means of personal expression if you ignore the contents in that way?

If we take the three main functional attributes of a human maker (heart, head and hand), that is, emotional response, intellectual planning, and making skills, which govern three of the above listed factors or aspects in making a book work, we might recognise which of them predominates in a given work, and so learn something about tendencies in the maker.

All three parts of the human psycho-physiological make up come into play in the conception and making of anything (even in carving a simple stick there has to be intention and direction to the carving; the carving is organised by visualising an outcome).

There may, however, be a predominance in an individual of one or more of the three main faculties. Depending on the predominant faculty which the maker habitually uses, determined to a great extent by psychological type, the ratio of the three could be apparent in the work displayed. Is the maker predominantly intellectual, emotional or physical in approach to a given situation? Different disciplines require the exercise of particular faculties. The great cultivators of civilisations: Science, Philosophy, Art and Religion, exercise different combinations of these faculties. Craft, Design, and Art do so too.

The idea of works created by the heart, head and hand has been of interest to this writer since beginning to study the arts in the late 1940s. Since then they have acted as a guide or framework in judging the constituents of a particular work and their integration as a complete whole. It is not therefore sufficient to have great manipulative skill; one has to be able to organise and control this in designing what is to be made with the skill, and the designing has to be informed with 'meaning', visual, verbal or tactile 'meaning'. Not only should the elements of visual art and design work in space or on a surface, but what *informs* the designed object is the ingredient that 'says' something not only about the response to the visual/verbal content of the text but also on such questions as man's place in the world, about his condition, about the consequences of his actions, which are usually portrayed by great writers, thinkers and poets. This *content* is the grist in the mill of the '*book-art-maker*'. Is the response of the viewer/ handler to the visual object on a similar wavelength to their response in contemplating the text? If it is then the maker has achieved some degree of success; but only that person who can handle, read and assess visual symbolism or metaphor, stated in colour and spatial forms, will know this by experience. The maker also needs to be in control of all these disciplines, then something beyond his conscious control may arise from a creative source deep in her or his being.

This 'something' I believe is innate in everyone. It is a matter of uncovering it, removing the impediments which bury it. Training of, and practice with, the hand encourages manual skills; training and practice in arranging colours and drawing shapes or forms encourages and hones designing skills; training and practice by reading, by looking at other arts and crafts can encourage conceptualisation. Conceptualisation on a high level demands direct training of the mind, the giving up of mental static in the form of recurring thoughts on the past or on the future to encourage stillness of mind and clarity of vision in the present. This has traditionally been achieved by such practices as meditation.

Each faculty on its own is without the means to manifest something coherently. All three need the support of the others and in making something all are involved in some proportion or degree. Fine arts require more of the conceptualising, emotional and intuitive characteristics (a balance of head and heart and hand weighted on the emotional aspect). A craftsman who cannot organise materials in a coherent way (that is, functionally design the object) is like a stool with two legs, or a bird with one wing. A maker may be the most manually skilled in the world on that level but, without idea or ability to harmonise and make the best use of his skill and materials, may produce ugly forms. A great designer without manual dexterity and co-ordination of his motor functions will have to depend on a skilled maker to manifest his mental creations. The atelier system in France was originated to do this for bookbinding designers, a division such as is used for industrial production of all kinds. For the independent British book maker this was seen as insufficient, as it meant that the specialist French designer was not experiencing the creative feedback from materials, although it did increase his production and turnover, possibly by a factor of ten.

The new ways in which leather can be manipulated are characteristic of much British work today. We use up offcuts, parings and other formerly discarded parts of skins. The new ways in which books can be presented as art objects are also largely a British invention; including multiples, sculptural containers, etc. As soon as a potential is discovered by one, the example gives rise to a plethora of experimentation by others. Innovation and invention with regard to the modern book is characteristic of British bookbinders today. This potential of the book as an art object has attracted several new young book makers whose work is included in the exhibition. New stars have appeared in the wings and are rapidly moving to the centre of the stage. Because there are few schools where bookbinding is taught, some are untrained in earlier finishing techniques and therefore invent their own, improvising with textiles, paper, leather, metals, wood and found objects.

The majority of designer bookbinders everywhere are not expected to work beyond the acquirement of the skills required to make decorated bindings, usually of an abstract graphic appearance and made with a high degree of skill. Many young binders starting off their careers are obliged also to take on jobbing bookbinding, or perhaps the repair of damaged books for booksellers and libraries. This is all very necessary, and so is the creation of special bindings for valuable early books, for presentations and all the trade bindings necessary for government, for the legal profession etc. where artistic concepts would be considered inappropriate. This is not, however, the concern of the experienced designer bookbinder or book artist except where conservation binding provides a main income.

Amongst our group there are some with artistic ambitions who aim at a more holistic approach. They give primary consideration to the text, and also how the structure of the book will work with this, and finally how to transmit the deep emotional impact which the author has had upon them to the public. They enjoy the manipulation of materials and engage in the book's total manifestation as an art object. Skill is not applied for its own sake, but to generate a response in the person who handles, sees, feels and thinks about the object to which they are bringing their own attention. As spin off from the controlled activity of the maker there may awaken a conscious intention to develop a harmony within him or herself between head, heart and hand, to achieve a right balance between senses, mind and spirit. This is the new spiritual dimension which has developed from an understanding of what it means to integrate all three functions of the human being and to transcend them. The act of making is a jumping off point from which to approach this development. Aiming in the direction of all round improvement is the cause of any positive result in this behaviour towards others and to the ecology of the planet. The aim of the fully focused and integrated maker must be a well-balanced person. This attitude, for any who hold it, may be spreading amongst British book makers, and also distinguishes them at present from most book makers elsewhere as far as our evidence seems to show[4].

There is a growing understanding that the book can be more than the sum of its parts. Perhaps this new impulse is some kind of compensation for all the threats, real or imaginary, on the imminent demise of the book due to the new technological alternatives in disseminating the electronic or printed word. Present developments in technology give the "*book-art-maker*" a wonderful opportunity to explore this. It should be pointed out to those who believe the book will disappear 'in ten years time' that 'hard copy' book production is actually on the increase.

It is to be hoped that something of the hidden values discussed above will come over in looking at these works, and that this introduction will give some pointers as to how one could be looking at them. There is much more to this new art than meets the eye, much more than craft skill and design organisation. There has to be a virtual meeting of minds in experiencing this form of art.

Philip Smith

Notes

[1] *When the term 'bookbinding' is used this should be understood simply as the physical part of the holding together of the disparate elements (printed text and covers) making up what we know as book. For some of us this is equivalent to the 'canvas stretching' preparation for a painting. Implicit in this term today – until a simple alternative designation is adopted for it as an art-expressive medium – are all the designing aspects (structural and aesthetic) and the artistic concepts (if any inform particular works). At present there is no term which adequately describes an artist developing his or her work beyond 'designer bookbinding'. I suggest* book-art-maker *is as near specific as possible a description for a lay person who visualises the term 'book artist' as being applied to an illustrator, or 'bookbinder' imagined to be producing conventional Victorian pastiches with 5-band gilded spines or the rows of brown calf bindings seen in stately houses.*

[2] The Canadian Bookbinders and Book Artists Guild *have foreseen in their title an alliance between two or three approaches to the physical book arts. The name of* The Bookworkers Guild *of the USA could also suggest the inclusion of several book arts. Book work in the Czech Republic has also become more art conscious during the past two decades.*

[3] *The making of artists' books in the USA is particularly widespread and prolific, and the practitioners concentrate on the interior art and paper book forms rather than design based on craft bookbinding. As yet too few are developing from the latter discipline, which also has its innovators in North America. The aims of makers of artists' books are different from those of bookbinding-based art; for them the text is the art ingredient, and with all kinds of social, political and artistic concerns.*

[4] *The increase in the international activity and exhibitions of hand-made books shows that the book is being celebrated as never before, but on a different level. For hundreds of years of its history the covers of the hand-made book were treated simply as a protection to the text-block, or as a decorative surface without further acknowledgement of the specifics of the work within. This traditional outlook continues in countries such as Germany and France where the psychological or spiritual potentials of book-art-making are rarely, if ever, considered, again except by makers of artists' books.*

The bindings reproduced in this catalogue are shown at scale 1:3
unless otherwise stated.

The Bindings

1 Glenn Bartley

Brave New World by Aldous Huxley
The Limited Editions Club, New York 1974
Gravures by Sara McAfee
270 x 198 mm ($10^{1}/_{2}$ x $7^{3}/_{4}$ in)
Bound 2000
Goatskin with airbrushed calf; goatskin and airbrushed calf doublures; gilt edges; gold and blind tooling.

2 Glenn Bartley

The Poems of Alfred, Lord Tennyson
selected by John D. Rosenberg
The Limited Editions Club, Cambridge 1974
Wood Engravings by Reynolds Stone
278 x 171 mm (11 x $6^{3}/_{4}$ in)
Bound 2000
Goatskin; paper doublures; coloured edges; gold and blind tooling.

3 Glenn Bartley

Seven Pillars of Wisdom, the Complete 1922 Text by T. E. Lawrence
Castle Hill Press 1997, in two volumes
288 x 210 mm (11½ x 8¼ in)
Bound 1998
Goatskin with airbrushed calf; goatskin doublures; gilt edges; gold and blind tooling.

4 Stuart Brockman

The Rime of the Ancient Mariner by Samuel Taylor Coleridge
Douglas Cleverdon 1929 (Fanfare Press London)
Engravings by David Jones
324 x 260 mm ($12\frac{3}{4}$ x $10\frac{1}{4}$ in)
Bound 2000
Goatskin with airbrushed calf; vellum doublures; gold and blind tooling.

An Introduction to the Rime of the Ancient Mariner by David Jones
Douglas Cleverdon 1972 (Will & Sebastian Carter)
Illustrated by David Jones
311 x 255 mm ($12\frac{1}{4}$ x 10 in)
Bound 2000
Transparent vellum over hand-dyed paper.

5 Stuart Brockman

The Searcher by Velona Pilcher
William Heinemann Ltd. 1929 (The Windmill Press)
Illustrated by Blair Hughes-Stanton
258 x 194 mm ($10\frac{1}{4}$ x $7\frac{1}{2}$ in)
Bound 1999
Goatskin, airbrushing; calf doublures; gold tooling.

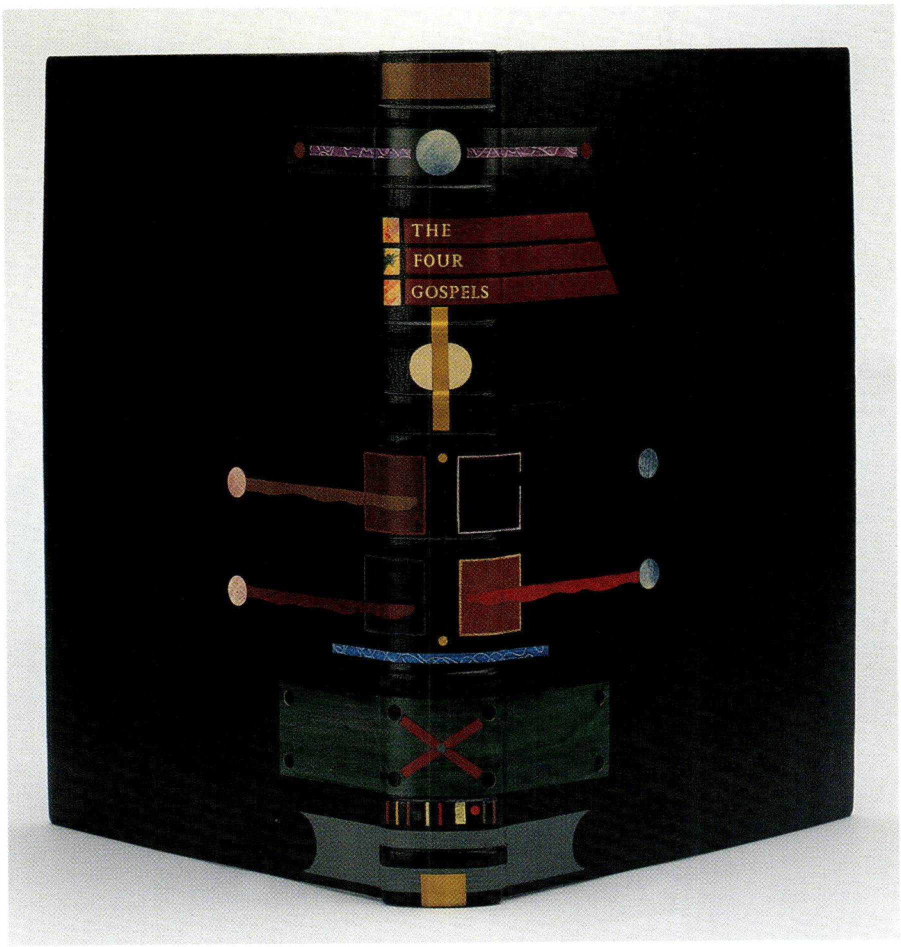

6 Lester Capon

The Four Gospels
September Press 1987, Limited edition
Illustrated by Eric Gill
342 x 247 mm ($13\frac{1}{2}$ x $9\frac{1}{2}$ in)
Bound 2000
Goatskin with stained calf, goat and paper; graphite top edge; gold and blind tooling.

7 Jeff Clements

Agamemnon by Eschyles, translated into French by Alexis Pierron
Abram Krol, Paris 1965, copy 1 of 76
Etchings and woodcuts by Abram Krol, signed
445 X 330 mm ($17\frac{1}{2}$ x $13\frac{1}{4}$ in)
Bound 1994
Goatskin with tooling in black.

8 Jeff Clements

The Symposium of Plato, translated by Tom Griffith
Libanus Press 1986, copy 75 of 355
Wood engravings by Peter Forster
275 x 185 mm ($10^3/_4$ x $7^1/_2$ in)
Bound 1993
Goatskin, salmon skin, acrylic paint; gilt edges.

9 Jeff Clements

The Homeric Hymn to Aphrodite,
translated by F. L. Lucas
The Golden Cockerel Press 1943
(Chiswick Press, London) copy 392 of 750
Engravings by Mark Severin
315 x 195 mm ($12^1/_2$ x $7^1/_2$ in)
Bound 2000
Goatskin, salmon skin, reptile suede, acrylic paint;
goatskin doublures.

10 Mark Cockram

The Collected Works of John Drinkwater
Sidgwick & Jackson Ltd. 1923 (Turnbull and Spears),
in two volumes
Copy 119 of 230, signed
230 x 171 mm (9 x 6 3/4 in)
Bound 1999
Goatskin with lacquered egg shell.

11 Stephen Conway

The Poems and Sonnets of William Shakespeare, edited by Gwyn Jones
Golden Cockerel Press 1960
323 x 204 mm (12¾ x 8 in)
Bound 1999
Goatskin, paper collage, studs.

12 Stephen Conway

The Story of Heledd, edited by Jenny Rowland
Gwasg Gregynog 1994
Engravings by Harry Brockway
335 x 210 mm (13 1/4 x 8 1/4 in)
Bound 2000
Goatskin with painted paper and mosaic.

13 Stephen Conway

All My Heroes Sang the Blues by Stephen Conway
Limited edition of 5 including CD, Incline Press 1999
183 x 197 mm ($7\,^1/_4$ x $7\,^3/_4$ in)
Bound 1999
Calf with painted paper and craquelle dyed calf.

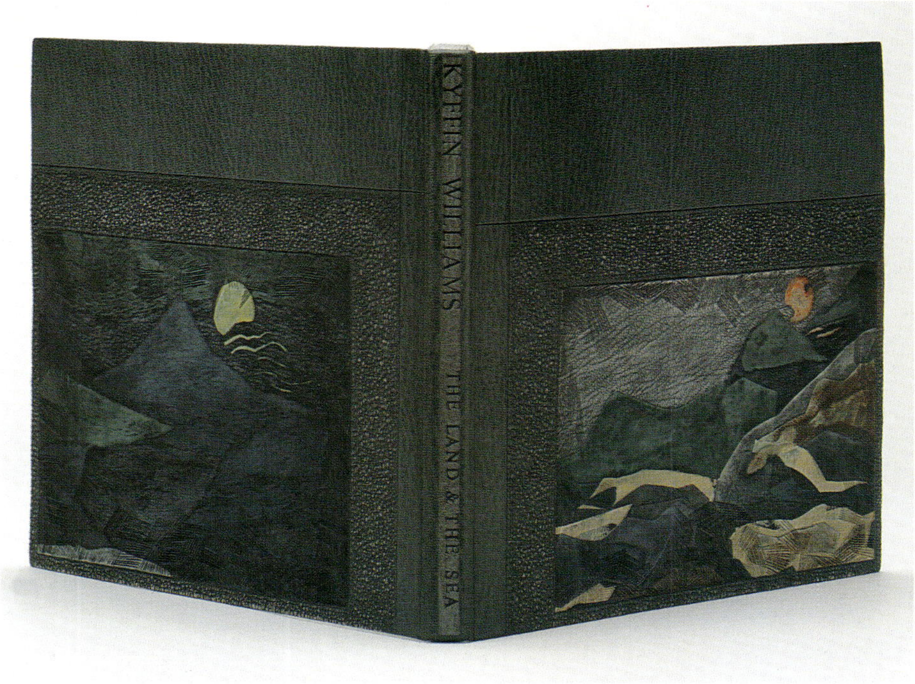

14 Paul Delrue

The Land and The Sea by Kyffin Williams
Gomer Press 1998
Illustrated by Kyffin Williams
238 x 270 mm (9 1/2 x 10 1/2 in)
Bound 1999
Sanded goatskin.

15 Paul Delrue

Wrenching Times – Poems from "Drum Taps"
by Walt Whitman
Gwasg Gregynog 1991, copy 409 of 450
Wood engravings by Gaylord Schanilec
375 x 235 mm (14 x 9 in)
Bound 1999
Sanded goatskin.

16 Sue Doggett

Mapping Narratives II by Sue Doggett
Sue Doggett 1999
Illustrated by Sue Doggett
140 x 132 mm (5 1/2 x 5 in)
Bound 1999
Screen printed paper with etching, photocopy, ink acetate, pencil.

17 Sue Doggett

Unity Universe, An A - Z by Sue Doggett
Sue Doggett 1999
Illustrated by Sue Doggett
148 x 130 mm (5 3/4 x 5 in)
Bound 2000
Linen, goatskin, watercolour, acrylic, gold leaf.

18 Jenni Grey

The Bible Designed to be Read as Literature
edited by Ernest Sutherland Bates
Folio Society 1957, in two volumes
Illustrated with Old Master drawings
285 x 390 mm (11 1/4 x 15 1/4 in)
Bound 2000, shown at scale 1:4
Goatskin, burr walnut, veneer, onyx.

19 Jenni Grey

Shocks by Gordon Lester Massman
Abattoir Editions and Yellow Barn Press 1983
Illustrated by Paul Otero
280 x 200 mm (11 x 8 in)
Bound 1999
Goatskin, wood veneer, brass studs.

20 Jenni Grey

November Twenty Six Nineteen Hundred Sixty Three by Wendell Berry
Illustrated by Ben Shahn
195 x 245 mm (7 3/4 x 9 3/4 in)
Bound 1999
Goatskin, maple wood, brass studs.

21 Simon Haigh

James Scott and William Scott, Bookbinders
by J. H. Loudon
Scolar Press 1980
260 x 174 mm ($10^{1}/_{4}$ x 7 in)
Bound 1996
Goatskin, wood veneer, paper, sprinkled edges.

22 Lori Sauer

The Testament of Charlotte B
edited by Marius Kociejowski
Libanus Press 1988, copy 181 of 220
Illustrated by Richard Shirley Smith
260 x 177 mm ($10^{1}/_{4}$ x 7 in)
Bound 1999
Vellum, dyed calf, silver chain;
silver gilt top edge.

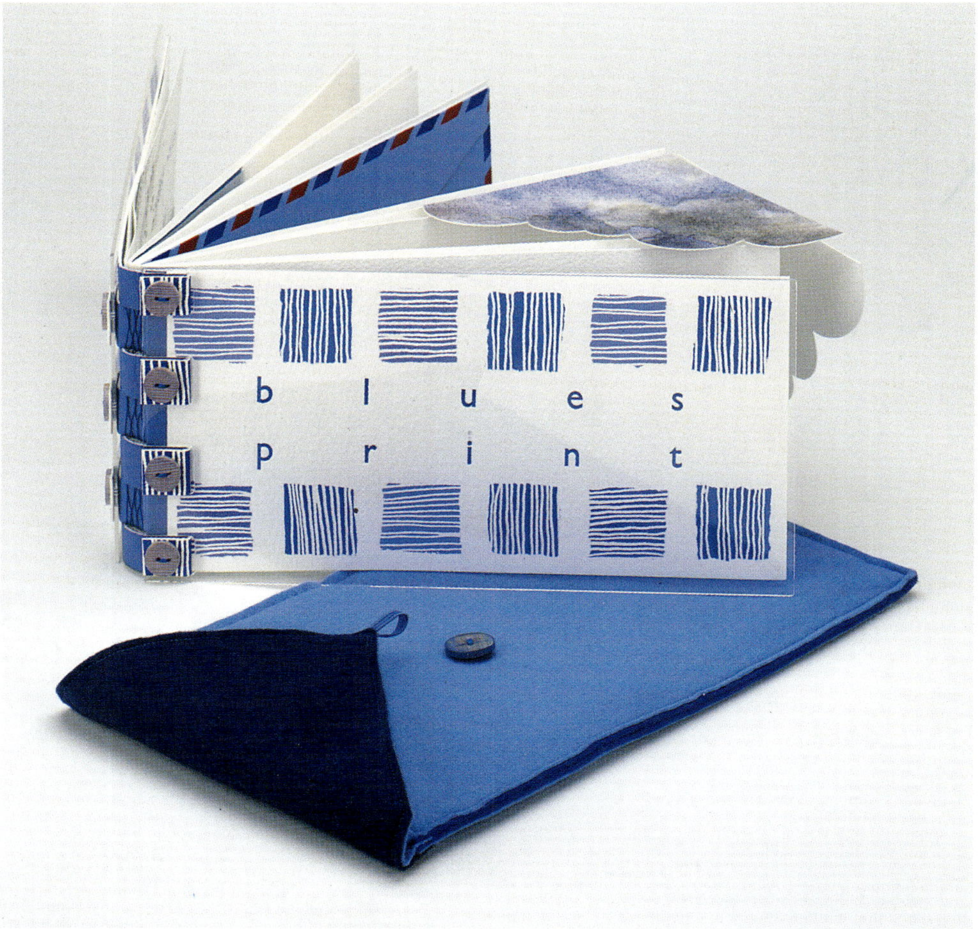

23 Angela James

Blues Print by Angela James
Angela James 1999, 1 of 2
126 x 254 mm (5 x 10 in)
Bound 1999
Paper, alum-tawed goatskin, wood, Perspex, acrylic ink.

24 Angela James

Of Gardens by Francis Bacon
The Fleece Press 1993
Illustrated by Betty Pennel
313 x 234 mm (12 1/2 x 9 in)
Bound 2000
Alum-tawed goatskin, wood, acrylic inks.

25 Angela James

The Garden and Other Poems by Andrew Marvell
The Folio Society 1993
Illustrated by Harry Brockway
112 x 80 mm (4 1/2 x 3 1/4 in)
Bound 1999
Alum-tawed sheepskin, calf, wood and maple veneer, enamel and acrylic paints.

26 Peter R. Jones

The Outing by Dylan Thomas
J. M. Dent & Sons Ltd., London 1978
Illustrated by Meg Stevens
130 x 167 mm (5 x 6 1/2 in)
Bound 2000
Goatskin and paper laminate, wooden dowel.

27 Peter R. Jones

The Red Badge of Courage by Stephen Crane
The Land Press 1988 (Gwasg Gregynog), copy 67 of 140
Illustrated with photographs from the Library of Congress, Washington D.C.
310 x 225 mm (12 1/4 x 9 in)
Bound 2000
Goatskin, wood, paper; goatskin doublures.

28 Midori Kunikata - Cockram

Setsugekka by Midori Kunikata
Midori Kunikata 1996
402 x 300 mm (15 3/4 x 11 3/4 in)
Bound 2000, shown at scale 1:6
Paper, silk, cloth, tape.

29 Mia Leijonstedt

Fifty Folio Epigrams
The Folio Society 1996
Engravings by Peter Forster
116 x 84 mm (5 x 3 1/2 in)
Bound 1997
Parchment paper, tin.

30 Trevor Jones

The Hue & Cry – A Tale of the Times by John Clare
Tern Press 1990
Illustrated by Nicholas Parry
280 x 190 mm (11 1/4 x 7 3/4 in)
Bound 2000
Goatskin, dyed calf; gilt top edge; blind tooling.

31 Trevor Jones

Urne Buriall and the Garden of Cyrus by Sir Thomas Browne
Cassell and Co. Ltd., London 1932
Illustrated by Paul Nash
321 x 232 mm (12 1/2 x 9 in)
Bound 1996
Dyed calf with goatskin and calf; gilt and palladium edges.

32 Flora Ginn

Recollections by Bernard C. Middleton
Bird & Bull Press 1995 (Henry Morris),
edition of 200
printer's copy, initialled
Twenty illustrations
257 x 170 mm (10 x $6^{3}/_{4}$ in)
Bound 2000
Goatskin; gold tooled goatskin doublures;
gold and black tooling.

33 Bernard C. Middleton

The New Testament
The Doves Press 1905
340 x 230 mm ($13^{1}/_{2}$ x 9 in)
Bound 2000
Goatskin; suede doublures; gold tooling.

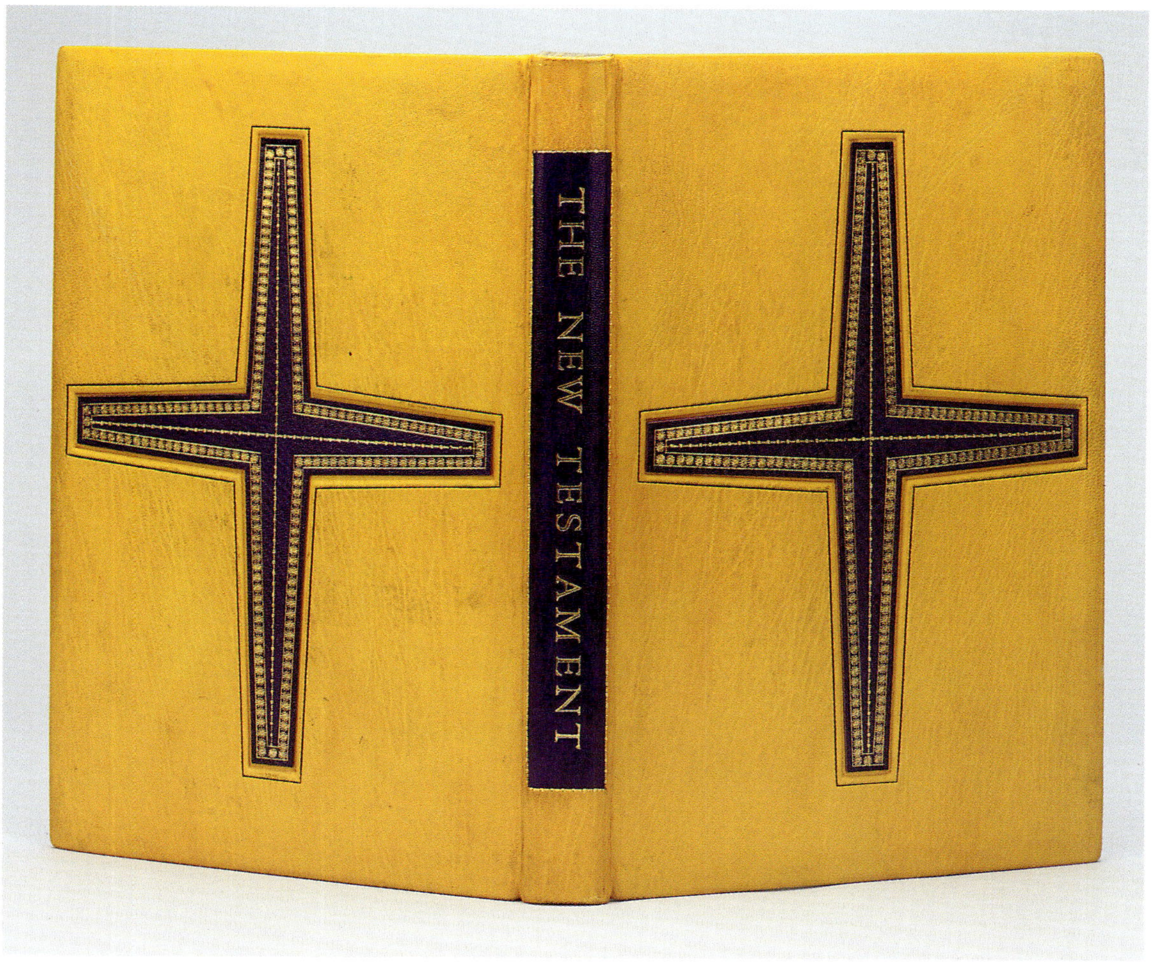

34 Ivor Robinson

King Lear by William Shakespeare
Ganymed Original Editions Ltd., London 1963 (Oxford University Press)
Limited edition of 275
Illustrations by Oscar Kokoschka printed from the stone by J. E. Wolfensberger, Zurich
463 x 346 mm (18x 13 1/2 in)
Bound 1994, shown at scale 1:4
Goatskin; sheepskin doublures; gold tooling.

35 David Sellars

The Waste Land by T. S. Eliot, edited by Valerie Eliot
A facsimile and transcript of the original drafts
including the annotations of Ezra Pound
Faber and Faber 1971, copy 117 of 500
292 x 228 mm (11 1/2 x 9 in)
Bound 1994
Sealskin, leather dust, painted vellum; suede doublures;
coloured tooling.

36 David Sellars

The Poems of Catullus,
translated by Charles Martin
Abattoir Editions 1979
(Harry Duncan and Mark Watson)
254 x 172 mm (10 x 6 3/4 in)
Bound 2000
Goatskin, vellum, ink; ink and watercolour
edges; coloured tooling.

37 David Sellars

Night - Faust's Study
by Johann Wolfgang von Goethe,
translated by Philip Wayne
Simon King Press 1995, copy 8 of 22
477 x 260 mm ($18^{3}/_{4}$ x $10^{1}/_{4}$ in)
Bound 2000
Goatskin, ink; transfer printed and graphite
edges; gold and coloured tooling.

38 Faith Shannon

'An Album' – a selection of bindings from 1982-98 by Faith Shannon
Faith Shannon 1998
215 x 173 mm (8 1/2 x 6 3/4 in)
Bound 1998, shown at scale 1:5
Card and paper laminate, reversed suede, printed paper, ink.

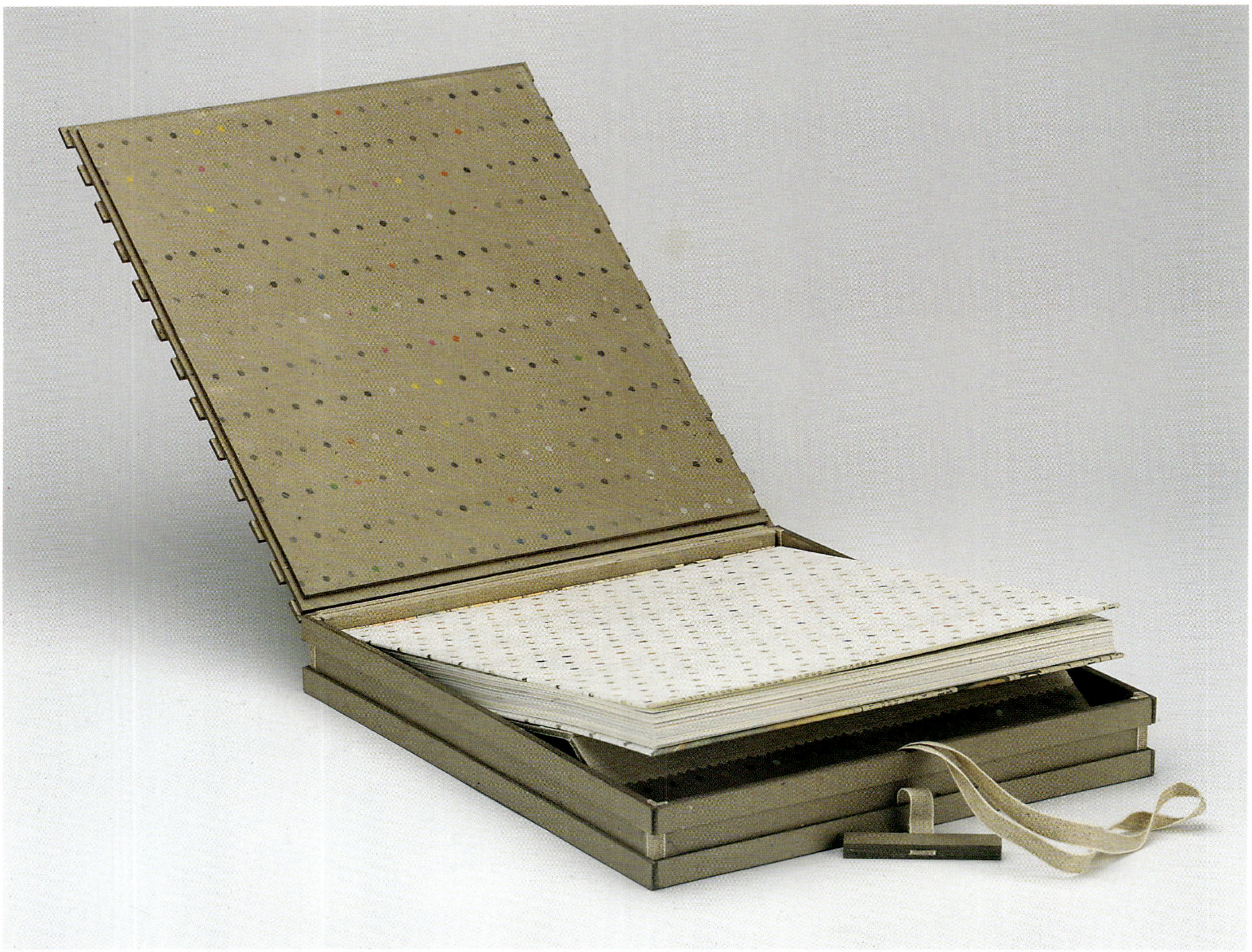

39 Faith Shannon

Paper Pleasures by Faith Shannon
French edition *Plaisirs de Papier*, Flammarion 1988
235 x 270 mm (9 1/4 x 11 in)
Bound 1995
Newspaper, computer paper.

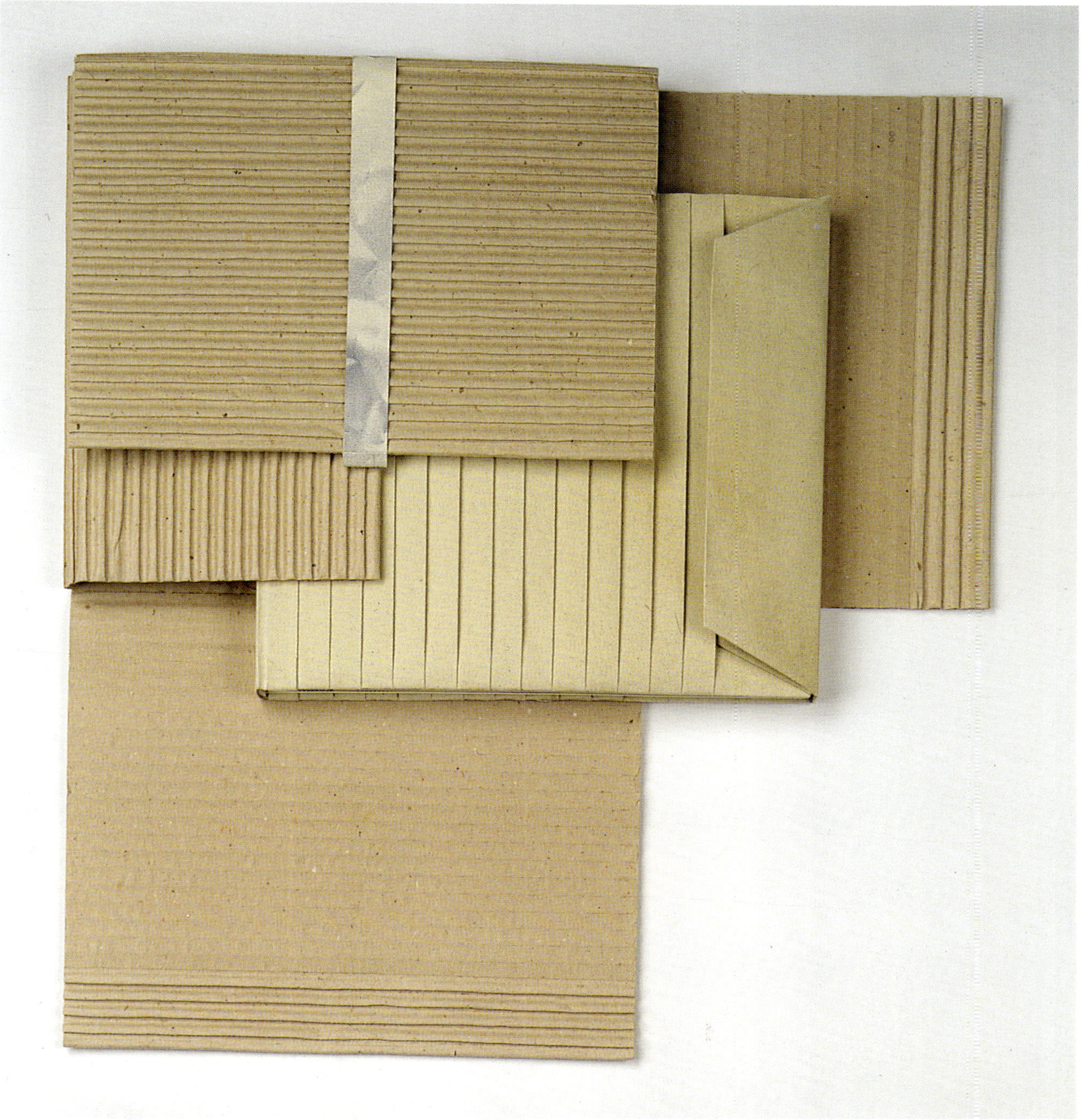

40 Faith Shannon

Paper Pleasures by Faith Shannon
Dutch edition *Veelzijdig Papier*, Gaade Uitgevers 1988
235 x 270 mm ($9\frac{1}{4}$ x 11 in)
Bound 1995
Paper, corrugated card.

VOL 1

VOL 2

VOL 3

41 Philip Smith

Inferno by Dante Alighieri
Talfourd Press 1983 (Tom Phillips),
in three volumes
Illustrated by Tom Phillips
429 x 334mm (17 x 13 in)
1324 x 354 mm (52 x 13 3/4 in) – height of 3 vol.
work in Perspex tower
Bound 1984 – 2000, shown at scale 1:4
Goatskin, 'maril'.

42 Philip Smith

Moby Dick by Herman Melville
Arion Press 1979 (Andrew Hoyem)
Wood engravings by Barry Moser
383 x 270 mm (15 x 10 3/4 in)
Content contiguous with sculptured container.
422 x 297 x 95 mm (16 1/2 x 11 3/4 x 3 3/4 in)
Bound 1989 – 1990, shown at scale 1:4
Goatskin, epoxy putty, balsa-wood.

43 Philip Smith

Six Fairy Tales from the Brothers Grimm, selected by David Hockney
Petersburg Press 1970 (Oxford University Press) Artist's proof copy no. XI edition B
455 x 325 mm (18 x 12 3/4 in)
Bound 1996 – 1997, shown at scale 1:4
Goatskin, 'maril'.

44 Sally Lou Smith

Songs from Shakespeare's Plays
by William Shakespeare
Officinae Bodoni 1974 (Giovanni Mardersteig)
270 x 177 mm ($10^{1}/_{2}$ x 7 in)
Bound 1995
Goatskin, acrylic paint; calf doublures; gold tooling.

45 Sally Lou Smith

Address of Paradise by Zofia Ilinska
Tabb House 1996
203 x 127 mm (8 x 5 in)
Bound 1999
Alum-tawed pigskin; goatskin doublures;
gilt edges; gold and blind tooling.

46 Julian Thomas

A Shropshire Lad by A. E. Housman
Tern Press 1990
Illustrated by Nicholas Parry
235 x 172 mm (10 x 7 in)
Bound 1999
Goatskin, vellum with dyed calf
and gilded leather, acrylic paint.

47 Julian Thomas

Wrenching Times – Poems from 'Drum Taps' by Walt Whitman
Gwasg Gregynog 1991
Illustrated by Gaylord Schanilec
375 x 235 mm (14 x 9 in)
Bound 2000
Goatskin, dyed calf, gilded leather; gold and black tooling.

The Binders

1, 2, 3 **Glenn Bartley** Fellow pages 16, 17

A childhood exposure to traditional finely bound books was the catalyst in choosing bookbinding as a career, but, it was only much later, when training at college and viewing exhibitions such as this, that I was made aware of the possibilities of using traditional bookbinding techniques in a more contemporary way.

My aim is to master, maintain and celebrate the many centuries of development in the craft skills that have been applied to hand bound books. Many binders in this medium have, over many years, largely eschewed these traditional methods of working, but I feel there are still possibilities to be explored.

Design elements are not making an individual statement as such, although I feel it is important to relate to the typography, design and theme of the text that, along with the book's protective container, create a unified whole. Also, the challenge for me is to produce special bindings that have a link with the past in their style and make up that still arouse those simple visual and tactile pleasures that create the 'warmth' of handling a well bound book.

These objectives will continue to stimulate my interest in the hand made book.

4, 5 **Stuart Brockman** Licentiate page 18

Bookbinding, in its present form, is the result of many centuries of experimentation and the development of complex interrelated techniques. It is the need to maintain, and attempt to match, the skills developed by previous generations, coupled with a desire for perfection, that drives my work.

In general my education was of a very analytical, technical and theoretical nature. My desire to use my more creative instincts and produce a high quality hand made product, with a need for functionality born from my engineering background, coupled with an enthusiasm instilled by my father's devotion to the craft, made bookbinding my natural future.

With my work I am aiming to combine successfully the "slick" tactile nature of continental bindings with the durability of the traditional British book. My ever-developing designs use traditional British forwarding techniques and established finishing methods (e.g. gold tooling) combined with very versatile modern methods (e.g. airbrushing). My designs are based on the interplay of form, line and depth, with individual elements crystallised into an abstract whole created from blocks of colour and precise shapes.

6 **Lester Capon** Fellow page 19

The design is centred around the spine and hinges of the book, integrating it with the underlying structure, just as Gill's illustrations integrate with the text. Rather than showing the structure, I am more interested in designing around it, and this binding is a continuation and extension of a previous work of mine incorporating the spine area, utilising treated and textured traditional materials.

7, 8, 9 **Jeff Clements** Fellow pages 20, 21

All three bindings are of classical writers, translated into English or French (*Agamemnon*). The construction of the bindings is straightforward, my attention focusing on the designs and materials.

The *Agamemnon* is a remarkable production, entirely by Abram Krol and includes strong linear etchings and woodcuts. I interpreted the story on the cover using minimal stark simplicity to portray the tragedy of the hero's homecoming and capture. The subtlety of line differentiation, free flowing with rectilinear, echoes the nature of the etchings and provides an insight into the story.

The *Symposium of Plato* uses the characteristics of argument and discussion as the point of departure and the design includes elements of formal logic contrasting with a strong diagonal axis to convey tension and the nature of debate. The colours were deliberately selected to indicate the delicacy of the main subject matter.

The most recent binding *The Homeric Hymn to Aphrodite* depicts a more extrovert attitude to sexual matters and the design here is concerned with the activities of the night and the provocative relationship between god and mortal. The combination of primary colours and animal skin accentuates this dramatic event.

10 **Mark Cockram** Licentiate page 22

I realise the book as a form of kinetic sculpture, construction being the foundation, allowing craftsmanship to permeate throughout. Design influences are varied from the organic to non-organic, manipulating materials, construction, colour and image to create an encapsulated environment that is in relationship with the text. Texture, both visual and tactile, plays a large role in the concept of my bindings. My challenge is to create an object/book that has form, function and life.

11, 12, 13 **Stephen Conway** Licentiate pages 23, 24, 25

I try to give my bindings a tactile quality, something I feel to be in keeping with the three-dimensional properties of the book format. Also, to give them an atmosphere, sympathetic to the text, through the use of colour, texture, materials and design. I am interested in allowing the container to become part of the work, giving the viewer a sense of involvement and anticipation. Another factor is the concept of producing the complete book through abstract/non-figurative art and music, and combining these elements to produce books that link design, sound and function.

14, 15 **Paul Delrue** Fellow page 26

I have always been most comfortable with leather, finding other materials unsympathetic. After a period during which colourful onlays were my first choice in design work, I began to experiment with 'leather pictures'. I evolved a technique that by sandpapering a patchwork of leather pieces I exposed a host of complementary and unexpected hues and shades. Further experiments revealed that by applying very dilute PVA and further sanding, an additional range of colours came to light. None of them was predictable from the original leather. This technique has become known as my "Lacunose" style.

16, 17 **Sue Doggett** Fellow page 27

"It's a poor sort of memory that works only backwards" (Lewis Carroll). Someone once said that "time is at the heart of all that is important to human beings" (Bernard D'Espagnat). It has occupied our thoughts since we ever became aware of its passing. These two books deal with the concepts of time and memory. *Unity Universe, An A-Z* consists of a collection of related quotations, imagery, signs and marks which I used/plundered/constructed as a way of exploring the concept, "time is just one damn thing after another" (Anon.). A way of marking time and holding the past in the present is through the souvenir. "Souvenirs allow one to be a tourist of one's own life" (Susan Stewart). *Mapping Narratives* is a collection of trinkets that tell tales about my life. It too follows the ritual gathering of text and image, scratching out and highlighting just as we do with our memories, constructing a history we can live with.

32 **Flora Ginn** Fellow page 38

This is my tribute for Bernard Middleton's collection of biographical anecdotes. The design represents his global reputation in the field of bookbinding. The lettering is repeated in mirror image as he reflects on his sixty years as a bookbinder and book restorer.

18, 19, 20 **Jenni Grey** Fellow pages 28 29, 30

Each of my bindings is a unique piece created specifically for the text, which provides the initial focus for my designs. My response to the text is personal, not attempting to illustrate or interpret the content, more to reflect on particular themes. The typography, illustrations, paper, size and structure of the book, all play a part in the design concept of individual bindings.

I employ a traditional approach to bookbinding design, applied in a contemporary way, utilising non-traditional material and constructions. I have developed a structure that allows me to build boards of wood, Perspex and etched metals, using leather only on the spine of the books where it is needed for its strength and flexibility.

Recently I have become more aware of how I would like my bindings to be viewed, and have extended my designs to the wooden boxes that display and protect the books.

My love of books is my incentive to preserve them through my bindings, which I hope will encourage others to value them. My aim is to produce tactile and visually pleasing work displaying a high level of craftsmanship, qualities intended to invite the reader to handle and use my bindings.

21 **Simon Haigh** Licentiate page 31

I have always enjoyed the feel of a well bound book.
In my binding I often use sculptured boards and raised onlays, to give the eye and the hand more to explore and enjoy.

23, 24, 25 **Angela James** Fellow pages 32, 33

The little book, *The Garden and Other Poems* by Andrew Marvell and the bigger book, *Of Gardens* by Francis Bacon, in a way, belong together. Both take gardens as their theme, of course, and although Bacon died five years after Marvell was born and their lives spanned very different periods of English history, these two editions were both printed in 1993.

The ideas for the designs came from the formality of Elizabethan and Jacobean gardens (Francis Bacon 1561 - 1626) and from stage sets of recent productions of Shakespeare. Both bindings use white goatskin or sheepskin for the covering material, with hand-painting and airbrushing. I wanted to convey a surreal image of the garden and its relation to nature and to man.

The Perspex binding may at first appear to bear no relation to the other two but it too uses painted cloud images inside although in all other respects it is very different. Sentences are printed in various shades of blue using letterpress, some pages having flaps to lift up or open out. The binding is a cross-structure one with an outer cover of Perspex through which the straps of the under-binding are slotted and fastened with handmade buttons. The under cover is a laminate of watercolour paper and white goatskin. The paper side is printed in blue and has onlays of resist coloured paper, the leather side is coloured with acrylic paint using a resist technique.

26, 27 **Peter R. Jones** Fellow page 34

Each binding is the result of its own evolution guided by a personal exploration of the medium of bookbinding. Every given text requires an individual approach. The structure of the binding will be determined by the physical nature of the text: size, format, layout, quality and weight of paper. Materials used will depend on the structure. Sometimes the combination of materials and structure are sufficient to represent my reaction to the text. Otherwise further elements will be introduced, often outlining an obscure reference to the content.

30, 31 **Trevor Jones** Fellow pages 36, 37

I have been binding now for forty-seven years, but am still stimulated by the many possibilities our craft provides for experiment and design, materials and structure. Each binding can be an adventure in a new direction, though with time we learn that some paths offer more than others. It took me twenty years to be at ease with the notion that my training in illustration could be of legitimate use for my bookbinding design.

John Clare's poem gives me the opportunity once again to seek inspiration from Eadweard Muybridge's historic photo resource *The Human Figure in Motion*. Many jokes are word-play. I like to subvert the conventions of bookbinding; here I have been playing jokes with a skewed version of the traditional half-binding format.

28 **Midori Kunikata – Cockram** Licentiate page 35

'The Book' is unique, as it is, for me at least, 'Total Art'.

The medium combines the skill of the papermaker, artist, printer and bookbinder to produce an object that is harmonious and self contained. Every aspect of the book has to be considered, from its very conception to choice of binding materials and eventual location.

Inspiration for my work stems from the text and illustrations, environment, material and physical texture. I feel that for my work to succeed all elements must be working in unison.

29 **Mia Leijonstedt** Licentiate page 35

In books I am fascinated by the relationship of the technical and the artistic. One does not restrict the other but rather feeds it and this is where I think the uniqueness of fine binding lies. For me, binding books is always both an adventure and a challenge. I am in constant dialogue with what I am binding; the book directs the work as much as I do. Sometimes it even takes on a life of its own. I am drawn to the tactile qualities of materials and prefer to choose books that inspire me in these terms, and through which the books' contents can be expressed. I am fond of paper, a material seemingly fragile yet incredibly versatile, and I often combine it with other organic components, always aiming to balance these with the functional requirements of the book's structure.

33 **Bernard C. Middleton** Fellow page 38

As is my wont, I have not attempted to do more than provide a pleasing visual effect with colour contrasts, shapes and the flash of gold.

34 **Ivor Robinson** Fellow page 39

I never 'think up' an idea for a binding. There will be a conscious and unconscious input into the act of drawing. The drawings will talk to me, and I talk to the drawings. A dialogue evolves on the paper. Challenged by a particular piece of work I may produce many drawings – perhaps combining aspects from several – and out of that will come something that I can use. The imagery may suggest a helmet, a slipping crown, lightning – but if one says '... here's lightning, a crown, a helmet', then what would that add up to? I am less interested in the ambiguities of symbolism than in the tangible relationships discernible between line, form and space. If my lines are able to convey anything at all in relation to 'King Lear', it might be a sense of that play's atmosphere and tension, but I would not wish to go further than that. The lines are first and foremost about drawing and about the act of impressing gold leaf into leather. Ultimately they are the story of their own making. They are themselves – and the text is their vehicle.

22 **Lori Sauer** Licentiate page 31

I enjoy experimenting with structure, technique and materials and I like to give myself a new challenge with each book. If the process of binding fails to teach me something I feel dissatisfied. The colours I choose usually come before the design and are as important; they are my response to the text's subject and tone. This binding uses the open joint structure and has vellum boards – both a first for me. The text is the 18th century tale of a young woman's abduction, captivity and eventual escape from her unwanted suitor, Mr H. She tells her story with restraint and propriety; she professes helplessness, yet devises and achieves her freedom. This combination of strength and vulnerability are what decided my choice of structure, materials and design.

35, 36, 37 **David Sellars** Fellow pages 40, 41

I began bookbinding as a cover for what was considered by my peers as excessive reading when a youngster. On starting creative bookbinding I naturally decided that ideas encompassed within a text were the central impulse and major focus of the medium. Occasionally a design comes complete during initial reading (*Faust*); usually research is necessary. Catullus needed Gibbon and Wiseman, *The Waste Land* was difficult, one was dealing with the drafts of the manuscript, birth and growth of the poem. The key was finally presented to me when teaching letterpress printing: a link between the galleys of type and the uniform grid of the modern city.

With such a multi-faceted physical presence as the book, other factors do influence the way a binding is arrived at visually. I am not interested in technique for its own sake, the structures, and palette of finishing activities are developed as a means to an end and the demand of a particular book. By chance there are a variety of structures: Catullus – tight joint, hollow back; *The Waste Land* – tongue-in-slot; *Faust* – a variant of tongue-in-slot.

Excitement, surprise and mystery are most important for me.

38, 39, 40 **Faith Shannon** Fellow pages 42, 43

Made for Contemporary Applied Arts, London, a series of exhibitions in 1998 commemorating their 50th anniversary, the curriculum vitae used for the covers of '*An Album*' has direct references (circled and underlined in red) to my links with the British Crafts Centre, as it was previously; it is partially obscured by rubbed printing ink over the stark, blatant surface. Designed to stand as a three-dimensional 'wheel' piece with the added flexibility of being a reference book, the detail sheets of each binding are simply produced – a mini portfolio. The slipcase has two versions in black album card or black vellum with a large bulldog clip to make it all a little less 'precious' and is strong enough to suspend the whole on a hook.

Although the bindings for *Paper Pleasures* may look subdued, they are intended to reflect some aspects of the book using apparently simple structures and everyday materials. Part of the philosophy of the book is to suggest noticing the often discarded materials around us. We do not always have to make a theatrical statement, but must always respect the making process appropriate to the book. Perhaps we should not take ourselves too seriously?

41, 42, 43 **Philip Smith** Fellow pages 44, 45

This group of four* book-works represents the four functions of head, heart, hand and 'soul' (animating principle) in portraying development towards a spiritual awakening in man. The works are combinations of the arts of literature, sculpture, painting and bookbinding. Physical book art requires a balance of the four disciplines along with the human functions controlling conceptual and imaginative techniques, in combination, to give it true definition.

The imagery for the *Inferno* represents the trials of the individual soul working through the unregenerate states of illusory being towards a potential Self-realisation, foretasting the continued search for divine being through the later Purgatorio and Paradisio cantos. The *Moby Dick* indicates the disastrous illusion of man's self-will in seeking vengeance against the forces of Nature. The *Grimm Fairy Tales* represents man's real search for the Truth of his Self, developing order out of chaos, and *The Four Gospels* shows man's relation to, and union with, the Creative Spirit throughout nature.

These works are included in order to represent "The Four Levels of Bookbinding", which is the subject of Philip Smith's visual lecture for "Bookbinding 2000" at the Rochester Institute of Technology in the U.S.A., and to demonstrate experimental sides of this modern medium.

* *The Four Gospels* was exhibited at Rochester Institute of Technology only, and is not illustrated.

44, 45 **Sally Lou Smith** Fellow page 46

I much prefer to bind a book with a substantial text to one with a negligible text or none. I read it, examine the typography and illustrations, if any, put it on the back burner and think about it occasionally. When the time comes for working on the binding, I know what the colours are going to be and how the design approach that I am currently exploring can be best adapted to this particular book. Without much conscious thought the book's final appearance has taken shape in my mind with only the details to be worked out. There is no rationale behind this method but it largely works for me, producing the binding design that would be appropriate to few other books.

46, 47 **Julian Thomas** Fellow page 47

When developing a design for a binding, my aim ideally is to capture the spirit of the book or at least to create a design that relates in some way to the book's content. Although I feel that the bookbinder must respond truthfully to his or her artistic and emotional convictions, I tend to avoid designs that may appear too obscure in relation to the text. I believe that an abstract design totally unrelated to the text is an evasive and self-indulgent response.

My bindings incorporate leathers that have been gilded, embossed and then sanded to reveal the colours beneath the gilding. The process is carried out before the leather is applied to the book by impressing the gilded leather using mainly the end grain of wood blocks, which have been sawn, naturally weathered or carved. I have been using this technique for the last three years and I am currently exploring its use in conjunction with gold tooling as a means of expression.

We would like to thank those who have given financial assistance towards the production of this catalogue:

Harmatan Leather Ltd.
J. Hewit & Sons Ltd.
Keith Adams
and to John Purcell Paper for donating the paper for the cover.

We would like to thank those who kindly lent bindings from their collections:

Lisa von Clemm: *Moby Dick* bound by Philip Smith

Mrs. V.M. Elliott: *Address of Paradise* bound by Sally Lou Smith

The Keatley Trust: *King Lear* bound by Ivor Robinson
Urne Buriall and the Garden of Cyrus bound by Trevor Jones
The Waste Land bound by David Sellars

Philip Smith: *Unity Universe, An A – Z* bound by Sue Doggett

Jeremy & Nicole Wilson: *Seven Pillars of Wisdom* bound by Glenn Bartley

We would also like to thank those who have given of their time and enthusiasm:

Keith Adams
Lester Bath
Flora Ginn
Nesta Davies
Kevin Howes

Designer Bookbinders
is the principal bookbinding society in Great Britain and one of the foremost in the world. It is devoted to the maintenance and improvement of standards of design and craft in hand bookbinding by means of exhibitions, teaching and publications.

Associate Membership is open to anyone anywhere in the world who is interested in bookbinding.

For further details and enquiries:

Designer Bookbinders
6 Queen Square
London WC1N 3AR
United Kingdom

info@designerbookbinders.org.uk
http://www.designerbookbinders.org.uk

Exhibition Committee: Glenn Bartley, Angela James

Selection Committee: Glenn Bartley, Flora Ginn,
Angela James, David Sellars

Catalogue: Lori Sauer

Photographs: Michael Barnett; Andrew Smart of
A. C. Cooper Ltd., London; Philip Smith

Published by Designer Bookbinders Publications Ltd.

ISBN 0 9538395 0 8

Printed and bound by
Smith Settle
Ilkley Road, Otley, West Yorkshire, LS21 3JP